Vertebrates

WHAT IS AN ANIMAL?

Ted O'Hare

Rourke

Publishing LLC

Vero Beach, Florida 32964

www.rourkepublishing.com

PHOTO CREDITS: All photos ©Lynn M. Stone

Title page: *The salamander is one of the vertebrates called amphibians. Frogs and toads are also amphibians.*

Editor: Frank Sloan

Cover and interior design by Nicola Stratford

Library of Congress Cataloging-in-Publication Data

O'Hare, Ted, 1961-
    Vertebrates / Ted O'Hare.
      p. cm. -- (What is an animal?)
    Includes bibliographical references and index.
    ISBN 1-59515-422-1 (hardcover)
   1. Vertebrates--Juvenile literature. I. Title. II. Series: O'Hare, Ted, 1961-   What is an animal?
    QL605.3.O39 2006
    596--dc22

Printed in the USA

CG/CG

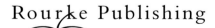

Rourke Publishing

www.rourkepublishing.com – sales@rourkepublishing.com
Post Office Box 3328, Vero Beach, FL 32964

1-800-394-7055

Table of Contents

Vertebrates

Vertebrates are animals with backbones and with skeletons inside their bodies, like us. The backbone is made up of sections called vertebras.

Vertebrate skeletons are made of bone and cartilage or just **cartilage**. Sharks and rays have cartilage skeletons.

Vertebrates have a **spinal cord** and a brain protected by bone or cartilage.

A dinosaur skeleton shows the vertebras of an ancient reptile.

Vertebrate Habits

Vertebrates survive in nature by using many special "plans." Each plan allows that animal to get the things it needs in special ways.

Vertebrates need food and water. They need to keep a body temperature that is not too low or too high. But reptiles, for example, don't eat, drink, or keep a comfy body temperature the same way that mammals do.

DID YOU KNOW?

Only about 3 of every 100 kinds of animals are vertebrates. The other 2 million or so kinds are boneless animals, or invertebrates.

The crocodile, like snakes and turtles, is a reptile. It warms its body by lying in the sun.

Kinds of Vertebrates

Scientists have placed vertebrates into five major groups called classes. They are fish, amphibians, reptiles, birds, and mammals.

A dog is a vertebrate, but it is also a mammal, just as we are. We mammals have some characteristics, like backbones, that we share with other vertebrates. But we are also different. We don't have feathers, for example, and fish don't have hair!

We and our dogs and cats are among the vertebrate animals known as mammals.

Where Vertebrates Live

Vertebrates of one kind or another live almost everywhere. They live in parks and cities as well as in forests, prairies, lakes, rivers, and oceans. A variety of birds and small mammals may live in your neighborhood.

DID YOU KNOW?
Certain kinds of fish live even in the deepest ocean trenches.

Some penguins live on or near the world's coldest land—Antarctica.

A northern elephant seal uses its flippers to scoop sand onto its back and avoid sunburn.

Amphibians spend their early lives like fish. As adults they can live on land as well as in water.

Vertebrate Bodies

Most vertebrates have two eyes and two pairs of **appendages**. People, for example, have arms and legs. Birds have legs and wings. Sea animals, such as whales and sharks, have flippers or fins.

You cannot see one of the important features of vertebrates. These are special cells. These cells help form many important **nerves** that invertebrates do not have.

Special nerve cells give many vertebrates keen senses, like this great gray owl's amazing eyesight.

Amazing Vertebrates

Other than insects, birds and bats are the only animals that can fly without help. Some birds fly very long distances each spring and fall.

Another amazing characteristic of some vertebrates is the ability to almost stop breathing and **hibernate**. Grizzly bears hibernate, as do ground squirrels, toads, and turtles.

Sockeye salmon on migration swim from the deep ocean into shallow freshwater streams.

Predator and Prey

Vertebrate hunters such as tigers, wolves, and weasels are predators. They kill other animals

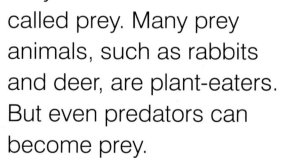

called prey. Many prey animals, such as rabbits and deer, are plant-eaters. But even predators can become prey.

18

Wolves are top predators, like bears and lions. They fear only other wolves.

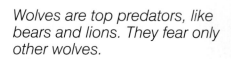

Baby Vertebrates

Most vertebrate babies are hatched from eggs. Some birds lay only one egg or two. Other vertebrates, like fish and amphibians, may lay hundreds of eggs. Nearly all mammals are born alive, without an egg around them.

Most vertebrate babies look like tiny copies of their parents. Baby amphibians, however, pass through a **larva** stage. At this stage they look and act more like fish than frogs, toads, and salamanders.

These hatching snow geese will soon dry their feathers and look like tiny adults.

People and Vertebrates

For us humans, vertebrates have been a source of fun, food, labor, sport, and companionship for thousands of years. We have even "improved" certain vertebrates, like cows and horses, by choosing the parents.

As the number of people grows, the natural world shrinks. Many kinds of vertebrates of land and sea will not survive without help from people—the most powerful vertebrates.

GLOSSARY

appendages (uh PEN dij uz) — things added to or reaching from, such as arms or legs

cartilage (KART uh lij) — a firm but often flexible material found in vertebrates; the material that supports the human nose and ears

hibernate (HI bur NAYT) — a period of long winter "sleep" when an animal's life support systems, such as breathing, slow down

larva (LAR vuh) — an early stage of growth for certain kinds of animals and in which an animal looks and acts differently than the adults of its kind

migrations (my GRAY shunz) — long-distance travels that take place at about the same time each year, usually in spring and autumn

nerves (NURVZ) — types of flesh that help receive and carry messages to the brain

spinal cord (SPYN ul KORD) — a stringlike system of nerves along an animal's spine (backbone); the nerve cord that carries messages to and from the brain

23

Index

Further Reading

Markle, Sandra. *Animal Predators: Wolves*. Lerner Publications, 2005
Pascoe, Elaine. *Animals with Backbones*. Powerkids Press, 2003
Silsbury, Louise and Richard Silsbury. *Classifying Reptiles*. Heinemann Library, 2003
Solway, Andrew. *Classifying Mammals*. Heinemann Library, 2003

Websites to Visit

http://www.biologybrowser.org
http://www.kidport.com/RefLib/Science/Animals/AnimalIndexV.htm

About the Author

Ted O'Hare is an author and editor of children's books. He divides his time between New York City and a home upstate.